Moritz Lehmann, Fabian Tatai, Markus Dietel

How to discover a new element? The synthetic creation of the yet hypothetical element ununennium.

GRIN Publishing

Bibliographic information published by the German National Library:

The German National Library lists this publication in the National Bibliography; detailed bibliographic data are available on the Internet at http://dnb.dnb.de .

Imprint:

Copyright © 2014 GRIN Verlag GmbH
Print and binding: Books on Demand GmbH, Norderstedt Germany
ISBN: 978-3-656-89582-4

This book at GRIN:

http://www.grin.com/en/e-book/289334/how-to-discover-a-new-element-the-synthetic-creation-of-the-yet-hypothetical

GRIN - Your knowledge has value

Since its foundation in 1998, GRIN has specialized in publishing academic texts by students, college teachers and other academics as e-book and printed book. The website www.grin.com is an ideal platform for presenting term papers, final papers, scientific essays, dissertations and specialist books.

Visit us on the internet:

http://www.grin.com/

http://www.facebook.com/grincom

http://www.twitter.com/grin_com

Moritz Lehmann, Fabian Tatai, Markus Dietel   *30.03.2014*

The synthetic creation of the yet hypothetical element ununennium

I. Experiment

$$\,^{208}_{82}Pb^{54+} + \,^{87}_{37}Rb \xrightarrow{\;-310,52MeV\;} \,^{295}_{119}Uue^{54+}$$

We direct a low-energy lead-208-ionbeam out of the LINAC 3 on rubidium-87-targets in order to create the new element ununennium with the atomic number 119. Either directly with the Lead Crystal Calorimeter or indirectly with the specific energy of the alpha particles that occur by the decay, we may prove the existence of single atoms.

Hopefully, this project doesn't take too much effort and an ion beam of such low kinetic energy will not make any problems in terms of radioprotection.

We decided to use the naturally present educts $\,^{208}_{82}Pb$ and $\,^{87}_{37}Rb$, whose sum of nucleons and protons equals exactly the numbers of $\,^{295}_{119}Uue$.

As projectile we utilize $\,^{208}_{82}Pb$, because all essential equipment for that isotope is already in place. The lighter weight of the target should cause no complications. The relative kinetic energy between the two cores is constant. A little hindrance could be that the metallic grid of the target lead would have smaller mashes as the one of rubidium, so that the collision's probability and the number of products could be lessened.

We assume a central collision of lead-ions with rubidium-atoms, which should transform moving to binding energy. The new ununennium-ion should therefore not be animated, but will be left with a rest momentum.

We need to calculate the speed necessary for a fusion of the lead-projectile with the rubidium target to the new element ununennium.

For this purpose we will first have to find out the part of the kinetic energy that is transformed into binding energy. With the conservation of momentum we detect the speed of the lead-ion as well as of the ununennium-ion.

The Bethe-Weizsäcker-formula is an accurate approximation of the mass of huge unknown cores.

The Bethe-Weizsäcker-formula[1] and its constants[1] for the calculation of the core intern binding energy are:

$$E_B = a_v \cdot A - a_O \cdot A^{\frac{2}{3}} - a_C \cdot \frac{Z^2}{A^{\frac{1}{3}}} - a_S \cdot \frac{(N-Z)^2}{4\cdot A} + \begin{cases} +a_P \cdot A^{-\frac{1}{2}} & for\ gg-cores \\ +0 & for\ ug-\ and\ \ gu-cores \\ -a_P \cdot A^{-\frac{1}{2}} & for\ uu-cores \end{cases}$$

$$a_v \approx 15,67MeV \quad a_O \approx 17,23MeV \quad a_C \approx 0,714MeV \quad a_S \approx 93,15MeV \quad a_P \approx 11,2MeV$$

A =Z+N means the amount of nucleons of the particular atom, Z the quantity of protons and N the quantity of the neutrons. Gg-cores are nuklei of atoms with even numbers of Z and N, for ug- or gu-cores only one of these variables is even, and for gg-cores none of these.

Using these values in the formula of binding energy gives these results:

$$A_{^{295}_{119}Uue} = 295 \quad Z_{^{295}_{119}Uue} = 119 \quad N_{^{295}_{119}Uue} = 176 \quad ug - core$$

$$E_{B_{^{295}_{119}Uue}} \approx 2083,77 MeV$$

Referring to this, the arisen 54 times positive ununennium has the mass[2] of:

$$m_{^{295}_{119}Uue^{54+}} \approx 274967,33 \tfrac{MeV}{c^2}$$

With lead-208 and rubidium-87 we use the known core masses for calculating the binding energy in order to get more accurate values. In this process the following relation to the proton[3], electron[4], and neutron[5] masses applies, if we assume that the core's mass equals the mass of the atom less the electrons:

$$m_p = 938,272046 \tfrac{MeV}{c^2} \quad m_e = 0,510998928 \tfrac{MeV}{c^2} \quad m_N = 939,565379 \tfrac{MeV}{c^2} \quad 1u = 931,494061 \tfrac{MeV}{c^2}$$

$$m_A = Z \cdot (m_p + m_e) + N \cdot m_N - \frac{E_B}{c^2}$$

$$E_B = (Z \cdot (m_p + m_e) + N \cdot m_N - m_A) \cdot c^2$$

The in u stated core masses have to be converted[6] into $\tfrac{MeV}{c^2}$.

These input parameters for lead-208[7] amount to:

$$A_{^{208}_{82}Pb} = 208 \quad Z_{^{208}_{82}Pb} = 82 \quad N_{^{208}_{82}Pb} = 126 \quad m_A = m_{^{208}_{82}Pb} = 207,9766521u$$

$$E_{B_{^{208}_{82}Pb}} \approx 1636,43 MeV$$

and with rubidium-87[8]:

$$A_{^{87}_{37}Rb} = 87 \quad Z_{^{87}_{37}Rb} = 37 \quad N_{^{87}_{37}Rb} = 50 \quad m_A = m_{^{87}_{37}Rb} = 86,909180527u$$

$$E_{B_{^{208}_{82}Pb}} \approx 757,86 MeV$$

The magnitude of difference between the core intern binding energies

$$\Delta E_B = E_{B_{^{295}_{119}Uue}} - \left(E_{B_{^{208}_{82}Pb}} + E_{B_{^{87}_{37}Rb}} \right) \approx -310,52 MeV$$

is simultaneously the amount of kinetic energy put into the fusion. For the reason of the endothermic process during the fusion the energy is negative.

$$E_{kin} = |\Delta E_B| = 310,52 MeV$$

The mass[4,6,7] of a lead-ion is circa

$$m_0 = m_{^{208}_{82}Pb^{54+}} \approx m_{^{208}_{82}Pb} - 54 \cdot m_{e^-} \approx 193701{,}42 \, \frac{MeV}{c^2},$$

if the already ionized lead-atoms are stripped of altogether 54 electrons after passing the carbon foil at the end of LINAC 3.

The light speed[10] in vacuum amounts to:

$$c = 299792458 \frac{m}{s}$$

A central, completely inelastic collision is present. The (in our case classic) conservation of momentum[14] applies:

$$m_{^{208}_{82}Pb^{54+}} \cdot v_{^{208}_{82}Pb^{54+}} + m_{^{87}_{37}Rb} \cdot v_{^{87}_{37}Rb} = m_{^{295}_{119}Uue^{54+}} \cdot v_{^{295}_{119}Uue^{54+}}$$

The rubidium-atom is reposing:

$$v_{^{87}_{37}Rb} = 0$$

$$m_{^{208}_{82}Pb^{54+}} \cdot v_{^{208}_{82}Pb^{54+}} = m_{^{295}_{119}Uue^{54+}} \cdot v_{^{295}_{119}Uue^{54+}}$$

The required energy E_{kin} is released during the inelastic burst:

$$E_{kin} = \frac{1}{2} \cdot m_{^{208}_{82}Pb^{54+}} \cdot v_{^{208}_{82}Pb^{54+}}^2 - \frac{1}{2} \cdot m_{^{295}_{119}Uue^{54+}} \cdot v_{^{295}_{119}Uue^{54+}}^2$$

With

$$v_{^{295}_{119}Uue^{54+}} = \frac{m_{^{208}_{82}Pb^{54+}} \cdot v_{^{208}_{82}Pb^{54+}}}{m_{^{295}_{119}Uue^{54+}}}$$

inserted the speed of the lead-ions can be calculated:

$$E_{kin} = \frac{1}{2} \cdot m_{^{208}_{82}Pb^{54+}} \cdot v_{^{208}_{82}Pb^{54+}}^2 - \frac{1}{2} \cdot m_{^{295}_{119}Uue^{54+}} \cdot \left(\frac{m_{^{208}_{82}Pb^{54+}} \cdot v_{^{208}_{82}Pb^{54+}}}{m_{^{295}_{119}Uue^{54+}}} \right)^2$$

Solved for the speed of the lead-ion the equation reads as in shown below:

$$v_{^{208}_{82}Pb^{54+}} = \pm \sqrt{ \frac{2 \cdot E_{kin}}{m_{^{208}_{82}Pb^{54+}} \cdot \left(1 - \frac{m_{^{208}_{82}Pb^{54+}}}{m_{^{295}_{119}Uue^{54+}}} \right)}} \approx 10{,}415\% \cdot c \approx 3{,}122 \cdot 10^7 \frac{m}{s}$$

Only the positive result is relevant for us.

We also solved the relativistic version of the equation system using the computer algebra system Maple:

Moritz Lehmann, Fabian Tatai, Markus Dietel   30.03.2014

$$\frac{m_{{}^{208}_{82}Pb^{54+}}}{\sqrt{1-\dfrac{v_{{}^{208}_{82}Pb^{54+}}^{2}}{c^{2}}}}\cdot v_{{}^{208}_{82}Pb^{54+}}=\frac{m_{{}^{295}_{119}Uue^{54+}}}{\sqrt{1-\dfrac{v_{{}^{295}_{119}Uue^{54+}}^{2}}{c^{2}}}}\cdot v_{{}^{295}_{119}Uue^{54+}}$$

$$E_{kin}=m_{{}^{208}_{82}Pb^{54+}}\cdot c^{2}\cdot\left(\frac{1}{\sqrt{1-\dfrac{v_{{}^{208}_{82}Pb^{54+}}^{2}}{c^{2}}}}-1\right)-m_{{}^{295}_{119}Uue^{54+}}\cdot c^{2}\cdot\left(\frac{1}{\sqrt{1-\dfrac{v_{{}^{295}_{119}Uue^{54+}}^{2}}{c^{2}}}}-1\right)$$

Maple does return the value

$$v_{{}^{208}_{82}Pb^{54+}}=10{,}41549059\%\cdot c\approx10{,}415\%\cdot c\approx3{,}122\cdot10^{7}\frac{m}{s},$$

which does not make much difference in terms of our accuracy.

Reaching this speed should already be possible with LEIR. The lead-ions will be concentrated in LEIR and PS and sent as dense packages onto the rubidium-targets in the test room T9.

The ununennium-ions that formed through the particle bombardment have the rest speed:

$$v_{{}^{295}_{119}Uue^{54+}}=\frac{m_{{}^{208}_{82}Pb^{54+}}\cdot v_{{}^{208}_{82}Pb^{54+}}}{m_{{}^{295}_{119}Uue^{54+}}}\approx7{,}337\%\cdot c\approx2{,}200\cdot10^{7}\frac{m}{s}.$$

2. Evidence

2.1 Direct evidence

If the lifetime is enough to fly a few meters out of the target, the ununennuim-ions could be proven directly with a mass spectrometer or the Lead Crystal Calorimeter. Different particles will be sorted out by a magnet or a Wien-filter.

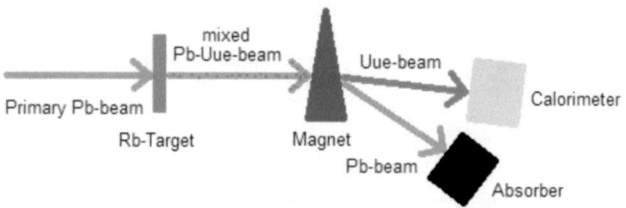

2.2 Evidence with decay

$${}^{295}_{119}Uue\xrightarrow{\;12{,}94MeV\;}{}^{291}_{117}Uus^{2-}+{}^{4}_{2}He^{2+}$$

Moritz Lehmann, Fabian Tatai, Markus Dietel

The ununennium-ions can also be stopped with an absorber shortly after the target. They withdraw electrons from their environment at once. We cannot prove the atoms themselves then. But because of their unstability, they as well as their daughter-cores emit particles like helium cores, positrons or neutrons, which have a certain kinetic energy depending on their origin-core. Exactly this information is provided by the available Delay Wire Chamber and the Lead Crystal Calorimeter. As a consequence we know their kind, kinetic energy as well as the position of the emitted particles and can prove the existence of ununennium.

If $^{295}_{119}Uue$ decays into $^{291}_{117}Uus$, which has the binding energy[1]

$$A_{^{291}_{117}Uus} = 291 \quad Z_{^{291}_{117}Uus} = 117 \quad N_{^{291}_{117}Uus} = 174 \quad ug-Kern$$

$$E_{B_{^{291}_{117}Uus}} \approx 2068{,}41 MeV$$

and therefore the mass[2]

$$m_{^{291}_{117}Uus} \approx 271253{,}58 \tfrac{MeV}{c^2} ,$$

and into $^{4}_{2}He^{2+}$ with the binding energy[11]

$$E_{B_{^{4}_{2}He^{2+}}} = 28{,}3007 MeV$$

and the mass[12]

$$m_0 = m_{^{4}_{2}He^{2+}} = 3727{,}379240 \tfrac{MeV}{c^2} = 6{,}644\ 656\ 20 \cdot 10^{-27} kg ,$$

the difference between the binding energies of educts to products

$$\Delta E_B = \left(E_{B_{^{291}_{117}Uus}} + E_{B_{^{4}_{2}He^{2+}}} \right) - E_{B_{^{295}_{119}Uue}} \approx 12{,}94 MeV$$

will be released in form of kinetic energy (converted into Joule[9]) :

$$E_{kin_{ges}} = \Delta E_B = 12{,}94 MeV \cdot 1{,}602 \cdot 10^{-13} \frac{J}{MeV} \approx 2{,}073 \cdot 10^{-12} J$$

Due to the conservation if momentum the mother-core and the daughter-core move outgoing from the mother-core in opposing directions. Therefore the following connection for the kinetic energy of the alpha particle[13] applies:

$$E_{kin_{^{4}_{2}He^{2+}}} = \frac{E_{kin_{ges}}}{1 + \dfrac{m_{^{4}_{2}He^{2+}}}{m_{^{291}_{117}Uus}}} \approx 2{,}045 \cdot 10^{-12} J$$

That is why the alpha particle for this collaps has a special speed[15] of

$$v = \sqrt{c^2 - \left(\frac{c}{\frac{E_{kin_{^4_2He^{2+}}}}{m_0 \cdot c^2}+1}\right)^2} \approx 2{,}475 \cdot 10^7 \, \frac{m}{s} = 8{,}254\% \cdot c$$

Reference: For a successful exothermic collaps ΔE_B has to be positive!

If the detectors capture a helium-core with this speed and measuring errors can be excluded with certainty, the existence of an $^{295}_{119}Uue$ -atom as well as of the arisen $^{291}_{117}Uus$ -atom is proven. But a helium-core can only be detected, if it was sent from the surface of the rubidium-foil and if the inner wall of the detector is appropriately thin, as alpha radiation can easily be screened out.

The calculation for the beta-plus-decay and neutron emissions would be similar.

Sources for the formulas, constants and values used in calculations:

1   https://www.uni-due.de/physik/fbphysik/Hauptseminar/WS0506/Ausarbeitung_Kernspaltung.pdf (15.03.2014)

     http://www.uni-tuebingen.de/faessler/Physik4/Physik42.pdf (15.03.2014)

2   Jürgen Eichler: Physik. Springer DE, 2011, S. 301.

     Ralf Steffler: Was die Welt zusammenhält. BoD – Books on Demand, 2001, S. 126 f.

3   http://de.wikipedia.org/wiki/Proton (15.03.2014)

4   http://de.wikipedia.org/wiki/Elektron (15.03.2014)

5   http://de.wikipedia.org/wiki/Neutron (15.03.2014)

6   http://de.wikipedia.org/wiki/Atomare_Masseneinheit (15.03.2014)

7   http://en.wikipedia.org/wiki/Isotopes_of_lead (22.03.2014)

8   http://en.wikipedia.org/wiki/Isotopes_of_rubidium (22.03.2014)

9   Hansruedi Schild, Thomas Dumm: Atom- und Kernphysik, Elektromagnetismus. Compendio Bildungsmedien AG, 2009, S. 49.

     http://de.wikipedia.org/wiki/Energie (23.03.2014)

10   Jürgen Eichler: Physik. Springer DE, 2011, S. 3.

11   http://en.wikipedia.org/wiki/Helium-4 (23.03.2014)

12   http://de.wikipedia.org/wiki/Alpha-Zerfall (23.03.2014)

13   Klaus Bethge, Gertrud Walter, Bernhard Wiedemann: Kernphysik: Eine Einführung. Springer DE, 2001, S. 227.

14   Ludwig Bergmann, Clemens Schaefe: Mechanik, Relativität, Wärme. Walter de Gruyter, 1998, S. 221.

15   Ferdinand Hermann-Rottmair, Prof. Detlef Hoche, Prof. Dr. habil. Lothar Meyer, Dr. Rainer Reichwald, Prof. Dr. habil. Oliver Schwarz: Physik Bayern Gymnasium 11. 1. Auflage, DUDEN PAETEC Schulbuchverlag, 2009, S. 107.

Moritz Lehmann, Fabian Tatai, Markus Dietel   30.03.2014